SKEPTIC

Once upon a time, I was a climate-change _____.

HUMANS

How could _____ affect this huge planet so much?

ACTIVISTS

Could _____ be creating a new cause to sell?

SCIENTISTS

Could _____ be trying to generate research grants?

COMPUTER MODELS

Could the _____ be wrong?

MEDIA

Could the _____ be over-hyping the science?

Though I was once a skeptic, I'm not one anymore.

ICE

The evidence is in the

DESPAIR

This knowledge of melting glaciers made me _____.

But _____ and defeat are not options.

OPTIMISM

We must invest in our _____ and in our strength.

FOR

This is the way .

EXTREME

With special thanks to EIS field staff: Jason Box, Dan Fagre,
Eran Hood, Svavar Jónatonsson, Adam LeWinter,
Jeff Orlowski, Tad Pfeffer, and Sport.

VANISHING GLACIERS AND CHANGING CLIMATE: A PROGRESS REPORT

ICE NOW

JAMES BALOG

FOCAL POINT

NATIONAL GEOGRAPHIC
WASHINGTON, D.C.

OOKING OUT THE WEST SIDE of my house, perched on a ridge in the Colorado Front Range, I can see the Arapaho Glacier, the most southerly glacier in the Rockies. The glacier hides in the shadow of the Continental Divide, below a 13,000-foot-high ridge of gneiss shot through with great veins of ivory-colored feldspar. The Arapaho isn't much of a glacier as glaciers go, just an oversized snowdrift, half a mile wide and no more than a couple hundred feet thick, with a few smallish crevasses to show the ice is actually flowing. By the end of summer, the inclined apron of pale gray ice at the lowest edge of the glacier, where the winter's snow has melted off, is striped. Spaced a dozen feet apart, the stripes mark the annual accumulation of snow from ancient winters, much like tree rings mark yearly arboreal growth. Photographs made a hundred years ago reveal that the Arapaho has shrunken dramatically, a process that is still ongoing. In the relatively few places where glaciers are found in the lower 48—the Sierra Nevadas of California; the Rockies of Colorado, Wyoming, and Montana; the Cascades of Washington and Oregon—the picture is the same. The ice is thinning and retreating as air temperatures rise and global precipitation patterns shift. (Seven small glaciers on northern California's Mount Shasta are a rare exception, as they benefit from recent warming of the Pacific Ocean.)

Shrinking glaciers are the canary in the global coal mine, the most visible, tangible manifestations of climate change on the planet today. In graduate school many years ago, I was trained in the geology of how the Earth's surface and, particularly, the mountains are shaped by the forces of tectonics, weather, and erosion. And as a mountaineer and nature photographer, I've visited many of the world's great mountain ranges and seen hundreds of glaciers. In spite of this background, I didn't understand how fast truly enormous quantities of ice could disappear until 2005, when I went to Iceland and shot a story for *The New Yorker* magazine. This led to a 2006 *National Geographic* magazine assignment to document changing glaciers in various parts of the Northern Hemisphere and South America. By the time this story was done (released in June 2007 as a cover story, "The Big Thaw"), the Extreme Ice Survey (EIS) was born.

EIS collects visual evidence of melting glaciers in Alaska, British Columbia, the Rocky Mountains, Bolivia, Greenland, Iceland, and the French and Swiss Alps. It is a collaboration between image-makers and scientists, all of us devoted to documenting the changes transforming Arctic and alpine landscapes today. We have approximately 27 (the number fluctuates depending on various technical

circumstances) time-lapse cameras shooting at roughly 18 locations. The cameras have been taking images of glacier and sea ice changes once an hour for every hour of daylight since spring 2007, and will continue doing so until at least autumn 2009, when nearly 400,000 frames will have been recorded. Our postproduction team compiles the individual frames into video animations, revealing how the ice is fluctuating. In addition, we are compiling a portfolio of still images in all the glacial regions we cover, shooting annual "repeat" photography at one-year intervals from fixed positions, and producing a documentary film about the ice and our experiences in the field.

Picking the time-lapse camera sites required careful study. The EIS team includes internationally distinguished glaciologists; they, and other specialists, suggest locations that provide a good representation of regional ice behavior. Camera sites are accessed on foot and horseback, by dogsled and skis, from fishing boats and helicopters. By the time the field team gets to the nuts and bolts of actually positioning our cameras, a wide range of variables, from travel logistics to the aesthetics of camera angles to the glaciers' sculptural qualities, have a bearing on where we shoot.

Engineering camera systems able to survive for two-and-a-half years and, in some cases, to withstand winds up to 160 mph and temperatures down to minus 40°F, has been a major undertaking. Our systems are based on ones used previously by glaciologists, including EIS team members Jason Box and Tad Pfeffer, but they are designed to shoot at a higher frequency over a much longer time period. Each Nikon D-200 camera is powered and controlled by a custom-made combination of solar panels, batteries, and other electronics. The cameras are protected from the elements by waterproof and dustproof Pelican cases, then mounted on sturdy Bogen tripod heads. All of the equipment is secured against the ferocious gales with a complex system of aluminum and steel anchors.

In our work, art meets science. Much like the Powell and Hayden expeditions to the American West during the 19th century, EIS is an adventurous, and sometimes dangerous, undertaking. But I believe the effort is worthwhile. Real-world visual evidence from our cameras has a unique ability to convey the reality and immediacy of global warming to a worldwide audience, to celebrate the otherworldly beauty of the ice-cloaked landscapes, and to help scientists better understand the mechanisms of glacial retreat. The mind-boggling pace of geologic change happening right here, right now, is something that I, as an observer of nature, as an Earth scientist, as a mountaineer, and as a citizen of the planet, feel compelled to document. EIS is a voice for landscapes that would have no voice unless we humans give them one.

There was a time when I wasn't quite sure climate change was real. I doubted it for the same reasons many people are skeptical: skepticism about the accuracy of the science and the scientists' computer models, and concern that the scientists, the media, and the activists were overstating what was happening. Most of all, I carried an assumption in my mind—inherited, I guess, from my cultural conditioning—that humanity couldn't possibly have the physical ability to change the very essence of our huge planet. I had certainly been seeing the year-to-year weather where I live get noticeably hotter and drier. And in the places I visited around the world during my photographic work, everyone seemed to be talking about how the climate wasn't what it used to be. But perhaps, I thought, we were all overreacting and the changes we were seeing were just natural variations.

When I learned more about the ice, I wasn't skeptical anymore. The most critical information is the temperature record from the Greenland and Antarctic ice sheets. The ice is built by steady annual accumulation of snowfall. At the bottom of the 2-mile-thick Greenland sheet, the ice is 230,000 years old; the bottom of Dome C in east Antarctica is an incredible 800,000 years old. Drillers, working from derricks not unlike those used to drill for water or oil, pull up 4-inch-diameter cores of this ice, in sections 39 inches long, one after another, until they hit bedrock. Glaciologists then make precise measurements of various oxygen isotopes preserved in the annual snowfall layers. The relationship among the isotopes gives a chronological record of temperature fluctuations through time. An essential note: This record is concrete physical evidence and is not a computer model.

You've heard, I'm sure, that the Earth is round and the sun rises in the east? Well, this geographic knowledge is as fundamental to the intellectual base of today's thinking people as I hope the story told by the ice cores will soon be. First, the ice shows that when the atmosphere contains more carbon, temperatures rise; when it contains less, the temperature cools off. Second, when the climate reaches a tipping point, it can flip-flop from dramatically colder to dramatically warmer regimes in as little as 1 to 3 years. Third, natural processes have made atmospheric carbon dioxide fluctuate between 180 and 285 parts per million by volume (ppmv) from 800,000 years ago to 250 years ago. In all that time, it has NEVER been above 285 ppmv. But, once people started burning copious quantities of coal during the industrial age that began at the end of the 18th century, atmospheric carbon dioxide steadily climbed above the built-in, natural limit of 285 ppmv. Today, the global average is 385 ppmv. In many urban areas the carbon dioxide level hovers near 500 ppmv. All of this is what those layers in the ice sheets tell us. This information changed me from being a climate change skeptic to a climate change believer.

If the story the ice is telling could be heard by everyone, there would no longer be any argument about whether or not humans are causing global warming. We are. Skeptics like to point out the natural variability that existed prior to A.D. 1750.

But they conveniently ignore the unmistakable anthropogenic, or human-caused, temperature rise since A.D. 1750. They even like to speak of climate regimes hundreds of millions of years ago, which is at best a convenient distraction: Continental positions back then were so radically different from today's, their climates are irrelevant to a discussion of modern climate change. In short, skeptical thinking is based on a limited understanding of Earth's history. And that's being charitable. In the case of certain lobbyists, consultants, and media commentators, I think the skepticism is based on political and financial bias that has nothing to do with the story glittering so clearly in the sunstruck ice.

The essays mixed in with the photos of this book are a precarious balancing act. The texts become cursory or unconvincing if I leave out too much detail to make them readable. On the other hand, they become wonky and ponderous if I put in too much detail. Most of the essays are thus a broad-brush look at important elements of the climate change discussion. A few introduce certain perceptual, psychological, or philosophical elements that are important but that haven't been much discussed. As the book goes to press, some of the issues mentioned in the essays—particularly the risk oil consumption poses to national security and our collective financial health—are going from being relatively esoteric concerns to being mainstream. The author in me worries about being outdated by the time the book comes out in early 2009. But the citizen in me would be ecstatic to be outdated; it would mean good things are happening in the United States.

Though the images and essays have a global message, much of it is specifically targeted to an American audience. That's where I live. That's where my greatest responsibilities are. Those are the people I can reach the easiest. But, more importantly, my countrymen are the ones who need to hear and heed the climate change story

the most. However you measure it—total carbon consumption or per-capita carbon consumption (see pages 114-115)—the United States of America has been the greatest carbon consumer. We have single-handedly invented the model of consumerism now being emulated globally, encouraging the whole world to throw more soot into the atmosphere. Having led the world there, we must now lead it to genuine solutions to climate change.

In spite of all the difficulties that climate change represents, I am optimistic. We are at a decisive turning point in the always-awkward dance between humans and nature. I must will myself to believe that we, the people, are now waking up and will do the right thing. In any case, as I say in the essay about heroes and fools, I am doing the best I can with the knowledge and skills I have. All the major currents of my adult life—mountaineering, geology, photography—have led to the Extreme Ice Survey. It is the only thing I should be doing at this time and place.

Think ——— ,think forward.

IXING CLIMATE change is an economic challenge. There's no doubt about it. It is also a technological and a political challenge. But considerable as these obstacles are, the money and tools and lawmaking can be handled. The biggest factors slowing us down are public psychology and perception. If we can see the problem of climate change more clearly and see it with greater imagination, we have a far better chance of actually doing something about it. Albert Einstein, in that wonderfully succinct way he had of framing issues, said it best: "Imagination is more important than knowledge. Knowledge is limited. Imagination encircles the world."

So what is holding us back? In the course of the past decade, as I have traveled hundreds of thousands of miles all around the Earth, I've noticed that most people in most places, including the United States, have the uneasy feeling that life on their home turf has changed. Temperature and moisture patterns are very different from what they were not so long ago and are certainly different from the weather measurements of the past century. What's causing the change? Is it natural variation? Is it man-made variation? If it's natural variation, we figure we can do nothing about it, which may be why so many are eager to stay with that explanation. This psychology of denial, allied with complacency, avoidance of responsibility, and fear, keeps us from taking action on climate change.

We try to deny the reality of climate change by claiming we don't have enough information yet, or we have a lot of conflicting information, or we are simply confused by the information. Complacency abounds. It's pretty comfortable, this lifestyle we have. Not thinking about whether or not this lifestyle might be self-sabotaging is much more convenient than acknowledging its foibles and doing something about them. Then there's an avoidance of responsibility: It isn't *our* fault we say, it's the fault of those rapidly industrializing countries on the other side of the world such as China and India. And finally there is fear, fear that changing business-as-usual will diminish the quality of our lives.

Let me note a quick, important aside about information in our society. Most journalistic outlets are obligated to give both sides of the story. Ninety percent of scientific specialists might agree on a given fact or trend in facts or theory, but the journalists feel an obligation to give approximately equal weight to the ten percent who don't agree. This balancing act is necessary to maintain the free and useful flow of information in society. But it leaves the knowledge stream far too easily polluted by information that is purposely misleading. We Americans are quick to call that kind of information "propaganda" when it flows through authoritarian states. We are less willing to recognize that our democracy is also inundated with propaganda; instead, we prefer softer terms like "spin," as in "spin doctor." The reality is that vested interests spend a fortune pumping propaganda through whatever journalistic pipelines they can find, countering solid, observed, physical, empirical facts with toxic effluent. And journalists, trying to produce a balanced story, are obligated to report both. This leaves the public so confused that it is unable to act, creating a perfect opportunity for denial.

Working singly or together, denial, complacency, avoidance of responsibility, and fear create individual and social paralysis. As long as our perceptions of climate change are flawed, we make no progress. Einstein had other words that apply here. Referring to the threat of nuclear annihilation that hung over us during the Cold War, he said, "We cannot solve our problems with the same thinking we used when we created them." We solve the problems by thinking smarter, thinking forward.

09.19.06 ▌FRENCH ALPS
Mer de Glace

A tourist walkway is a graphic indicator of the glacier's thinning. In 1988, the platform on the top right touched the glacier's surface. Over the next 18 years, officials had to add downward extensions of the walkway so that visitors could still reach the glacier.

PAGES 24–25:

09.25.06 ▌SWISS ALPS
Trift Glacier

The Trift has retreated nearly two miles since 1850; more than a quarter of this retreat occurred during the past ten years.

09.27.06 ▌SWITZERLAND
Rhône Glacier

Once a tourist attraction for its reach into the Rhône River, the Rhône Glacier may attract another kind of tourist in the future: boaters. The glacier is now melting and forming a lake at its base. Experts estimate that most of the glaciers in Switzerland, like this one, will be gone by A.D. 2100 if the melting rate continues at the current pace of three percent per year.

DESPAIR IS NOT AN OPTION.

OPTIMISM ABOUT THE ecological future of humanity doesn't necessarily come easily to me. Climate change is tied to some enormous geopolitical issues—energy supply, economics, technology, and population growth—and the human race can be impervious to common sense, so pessimism sometimes seems like the only refuge of a rational mind.

A glacier, however, showed me a different way of thinking. I spent ten days in April 2006 working at the terminus of the Sólheimajökull, a glacier in Iceland. I carefully selected and photographed a dozen different perspectives to which I would return six months later and record the ongoing retreat. On one typically drizzly day—if ever there was a day for Bergmanesque Scandinavian angst, this was it—I was perched high up on a rocky, wind-scoured knoll, working on a multiframe panorama along the east side of the ice. Below me, a line of tall gray stakes, spaced at intervals ranging from 100 to 300 feet apart, marched up the valley toward the point where the ice ended. They marked the steady recession, since 1999, of nearly half a mile.

It may not be obvious to nonphotographers, but photography is something of an act of love: The sustained attention we give to our subjects draws us closer and closer as we get to know them better. In this case, intense study of a relatively small area of ice caused me to become infatuated with the glacier's every shape and tone and nuance. I was filled with despair when I realized that the object of my fixation just might vanish before I returned in October. Further, I wondered, did the human race really have the intellectual and emotional fiber to stop changing the climate?

Possibly not.

Yet within a week or so, a different answer occurred to me: Defeat and despair are not acceptable.

With that realization, I ripped out the wiring of my ingrained pessimism and inserted a new circuit of optimism. Pessimism and despair are a self-fulfilling trap. President Franklin D. Roosevelt's famous quote, "The only thing we have to fear is fear itself," tapped into the same root of understanding: Negative thinking leads to negative consequences. As I squinted against the Icelandic mist that day at the Sólheimajökull, I found it

absolutely intolerable that the people of my time—I, my family, my friends, my working peers, my community, my state, my nation—would be the ones responsible for destroying something as monumental as the climate of this huge planet. The idea was too sickening to accept. I have also come to see defeatism as inherently self-absorbed; self-absorption wallows in the present instead of letting us see the distant horizon called the future. Whether our future is comfortable or miserable, honorable or degraded, will be dictated by the willpower and heroism we're willing to bring to the task of changing our ways.

Wishing our society to be strong-willed and heroic doesn't make us so. I understand that. But we all know that nothing great was ever achieved without passion, dedication, willpower, and determination. Call my wish a leap of faith if you will. So be it. Pessimism is a dull gray alternative. Optimism moves mountains.

04.13.06 ▮ ICELAND ▮ Sólheimajökull Glacier

09.16.07 ▮ ICELAND ▮ Sólheimajökull Glacier

The glacier retreated 245 feet during one melt season.

09.14.07 ▮ ICELAND
Jökulsárlón

Icebergs that originated in the vast expanse of the
Vatnajökull Glacier, the largest glacier in the country,
decay and melt in a tidal lagoon.

OPPOR

Today is an _____ ; tomorrow will be a crisis.

TIME IS SHORT. All the new global warming gases we're dumping into the atmosphere, whether they come in the form of carbon dioxide (CO_2), methane, or nitrous oxide, make the air thicker and denser every year. Carry on with business as usual and our atmospheric CO_2 will rise from its current average of 385 ppmv today to 500 ppmv by 2050. By 2100 it will be between 720 ppmv and 1020 ppmv. Average global temperatures are projected to rise 1.8°F to 2.7°F by 2050 and a total of 3.6°F to 6.3°F by 2100. After 2100, the warming could continue steadily, possibly with disastrous results (see pages 114-115).

What does the warming do? To varying degrees in different parts of the planet, it dries up water supplies, disrupts agriculture, increases prices of food and water, makes extremes of weather more extreme, raises sea level, sets off more wildfires that burn more intensely, spreads mosquito- and rodent-borne disease, causes the extinction of some animal and plant species, and helps undesirable animals and plants proliferate. These impacts are already showing up globally, including in North America. According to the best current estimates, sea level will rise at least 2 feet, probably 3 feet, and possibly 4 to 5 feet by 2100. The rate is more likely to accelerate over time than it is to stabilize or slow down. Other problems may not pan out exactly the way science is projecting; new problems we don't yet anticipate will show up. Whatever happens, the human race will have an interesting future.

Most serious students of climate change are convinced we have 10 to 20 years at most to begin making major changes in how we generate energy—and we must continue the process indefinitely after that. The day of no return will almost certainly come during the present century; after that the warming will be so far along, and the methods of carbon consumption so hopelessly entrenched, that arresting the damage will be impossible.

We can't be naïve. The job of changing our energy consumption habits and putting the climate back on track is enormous. The kind of technological resolve such change will take becomes commonplace once it is put into use, but step back from it and look at it with a fresh eye, and it is nothing short of miraculous. Satellites, manned space programs, lunar landings: these are all fabulous achievements, now taken for granted. What about America's mobilization for World War II, when President Roosevelt set the seemingly impossible goal of building 60,000 aircraft, but industry managed to build 230,000 instead? What about our 21st century ability to disassemble and reassemble the genetic code of life, or to transport information at warp-speed on the Internet? This kind of technological magic will be critical to our future. In any society willing to take on the challenge, new industries will be born and new jobs created.

Yes, there will be costs as the energy supply system adjusts: The definitive analysis, authored by Sir Nicholas Stern, former chief economist of the World Bank, calculates the cost of holding atmospheric carbon at roughly 550 ppmv to be one percent of global gross domestic product (GDP). Stern projects the long-term cost of doing nothing to be a minimum of 5 percent of global GDP forever, and possibly as much as 20 percent.

Unless we take action, today's opportunity will become tomorrow's crisis.

PAGES 36-37: Seawater has polished the surface of this iceberg in Iceland.

02.09.08 ▮ ICELAND
Jökulsárlón

On its way to the North Atlantic Ocean, an iceberg the size of a compact car washes up on a black sand beach under a starry sky.

PAGES 42-43:

02.09.08 ▮ ICELAND
Jökulsárlón

High tide brings an endless procession of ice fragments onto the beach. The "ice diamonds" are unique sculptures created when they tumble in the surf and onto the sand. They will vanish during the next high tide.

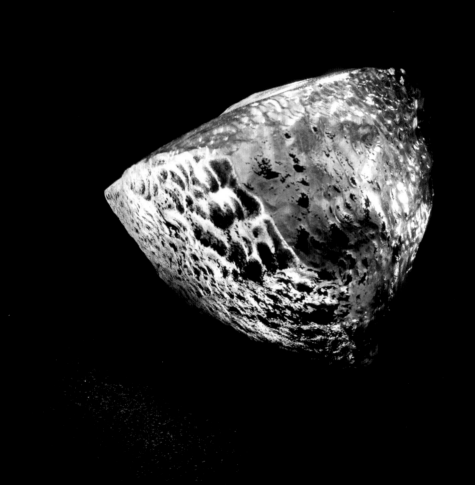

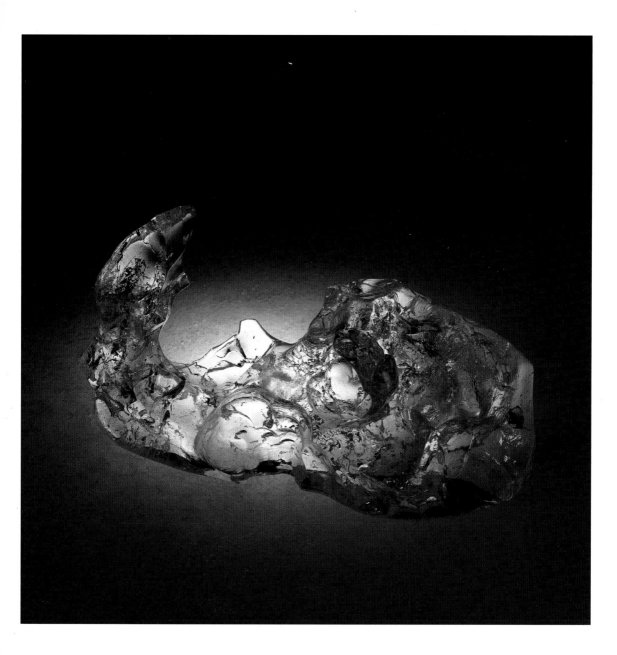

AMONG THE MOST INFLUential figures in the history of global oil markets was Sheikh Ahmed Zaki Yamani, the Saudi oil minister from 1962 to 1986. During a 2001 interview he said, "The Stone Age did not end for lack of stone, and the 'oil age' will end long before the world runs out of oil." At the time, he was expressing his dismay at $30 per barrel prices. His words were meant as a caution to petroleum producers, a warning to keep prices in check so oil consumers wouldn't be tempted to find a cheaper fuel. Today, prices are more than two to four times that amount. The concept of $30 oil has become as quaint as a camel caravan bearing frankincense and myrrh, since by all credible accounts, oil prices will never drop anywhere near $30 again. It is bizarre to find our society still unsure about whether or not we should find alternatives to oil.

So why did the Stone Age end? Or, more to the point, why would the oil age end? The answer is obvious: because people realize there is a better way. The human race has no God-given obligation to use up every molecule of petroleum before we move to other energy sources.

Technology is not something that happens outside human control. We create technology. We will it into being. It works for us; we don't work for it. When one tool doesn't work, whether it's a screwdriver or an oil tank, we can and should try another. (To be sure, we still use stones and bronze today, but their usefulness is limited. Similarly, we will always use petroleum products, but ideally not as much.)

Consider the story of whale oil. Through America's early colonial years, whale oil was the most desirable fuel for use in lanterns and candles and for lubricants. The internal combustion engine hadn't even been invented. Yankee whale men got rich selling whale oil to the world; their wealth played a major role in shaping the economy and society of America. But because they had over-hunted the cetaceans, the whalers had to sail ever farther from their home ports in New England. The price of whale oil rose. Gas produced from coal and kerosene derived from natural tar started to eat into markets previously owned by whale oil. Then, on August 27, 1859, a well drilled into the sleepy hills of Titusville, Pennsylvania, yielded a gusher of oil. Immediately, a rush for "black gold" was on. By the end of 1860, petroleum-based oil had taken over the market, the price of whale oil

had dropped, and the American whaling industry had collapsed.

Back to the present. Whether or not we've passed "peak oil" is irrelevant. Even though huge quantities of oil still lie buried in the Earth's crust, the limitations of oil as a fuel are now manifest. Our inability to end our petro-affair is akin to the emotional frustration bedeviling the breakup of so many romances: You know it must end because the relationship is driving you batty, yet you're afraid to leave because you don't know if you'll find anything better. You can't live with it and you can't live without it. But as your perception becomes more lucid, it becomes easier to end the relationship. Like unhealthy, addicted love, our affair with oil is causing us more pain than pleasure. Right now would be a terrific time to end the romance and run, not walk, toward greener pastures.

PAGES 44-45: Wind-driven snow peppers an "ice diamond" on the beach in Jökulsárlón, Iceland.

08.24.07 ▎ GREENLAND
Disko Bay

Icebergs 200 feet tall, formerly part of the Greenland
Ice Sheet, float into the North Atlantic Ocean, raising
sea levels as they melt.

03.15.08 ▌ GREENLAND
Ilulissat Isfjord

The sea surface downstream from the Ilulissat Isfjord's calving face is packed so tightly with icebergs small and large that you can walk many miles and never touch water.

PAGES 52–53:

07.18.06 ▌ GREENLAND
Greenland Ice Sheet

This meltwater lake was created by warm summer temperatures on the surface of the Greenland Ice Sheet, 30 miles south of the Ilulissat Isfjord. The water eventually flows into moulins, or stream channels, that drill their way down through the ice, then flows out the base of the glacier into the ocean. Meltwater lubricates the glacier bed and makes the ice flow faster into the sea. Global warming has caused melting to occur dozens of miles farther inland than 20 years ago.

OPPORTUNITY WILL MAKE "TECHNO SAPIENS" RICH.

AS ANY THINKING INDIVIDUAL surely has observed, technology has been both the blessing and curse of civilization. One minute it bestows unprecedented power on the human race: Think how easy online search engines have made it to collect information. The next minute, we are as helpless as newborn kittens when the tools we depend on fail:

Think how frustrating computer glitches can be and how difficult they are to fix.

During much of the 1990s, I produced a photo portfolio tracing the convergence of people and technology. I called it "Techno Sapiens." The photos show the human animal morphing into man-machine hybrids. The imagery certainly doesn't radiate the confidence one hears from the devotees of the wired world; instead it is something of a eulogy for the loss of the organic world. So given my skepticism about technology, I was surprised during one windswept day in a tent in Greenland to find myself writing in my notebook, "Techno Sapiens just might save us from climate change."

All around us, I realized, are "techie" visionaries. They see the necessity of changing how we generate electricity, transport ourselves, and keep our buildings warm. These visionaries are individuals, small corporations, large corporations, and even countries. They are mining their way through a mother lode of ideas, looking for the next gold strike to make them wealthy. Thirty years from now, the way we use and generate energy will be very different, even if most of us can't imagine that future. It has always been this way, with the visionaries out in front. What they imagine today will become tomorrow's commonplace.

In some obscure office or garage, a brilliant thinker or tinkerer—quite possibly a young one, as Bill Gates was when he dropped out of Harvard and started down the path that led to Microsoft—is developing a technology the rest of us haven't even dreamed about. Simultaneously, existing multibillion-dollar corporations are sifting through a global sluice box of ideas, actively trying out some new technologies while waiting for economic or regulatory circumstances to change before they try out others. Countries like Denmark, Israel, and Switzerland are already demonstrating spectacularly innovative new ways to transport people and power through their societies. At whatever scale they're operating, these visionaries are in the process of minting the next great fortunes. Thomas L. Friedman, the journalist, put it perfectly, "Green is the new gold." The rest of us can't quite see the gold flakes glittering through the carbon sludge in the bottom of the prospector's pan. But the "techie" visionaries can.

08.11.08 ▌BOLIVIA
Chacaltaya Glacier

This mountainside was once completely covered by a glacier. Remnants from the 1940s of what was once the world's highest ski area at nearly 17, 000 feet can still be seen on the ice. Experts estimate the glacier will be entirely gone in the next few years.

PAGES 58–59:

08.20.06 ▌BOLIVIA
Huayna Potosi

In the Andes mountain range, a mountaineer climbs through a surreal landscape of miniature "neve penitentes." The penitentes are created by melting processes that are thinning the region's glaciers 3 feet each year.

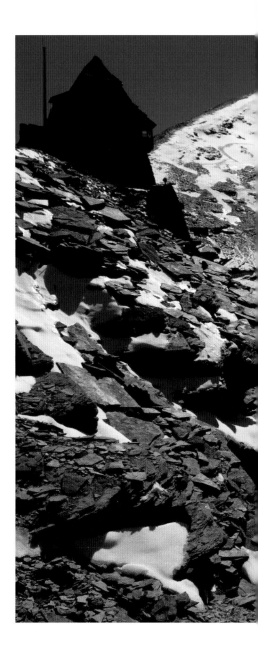

---- 08.11.2006

---- 08.11.2008

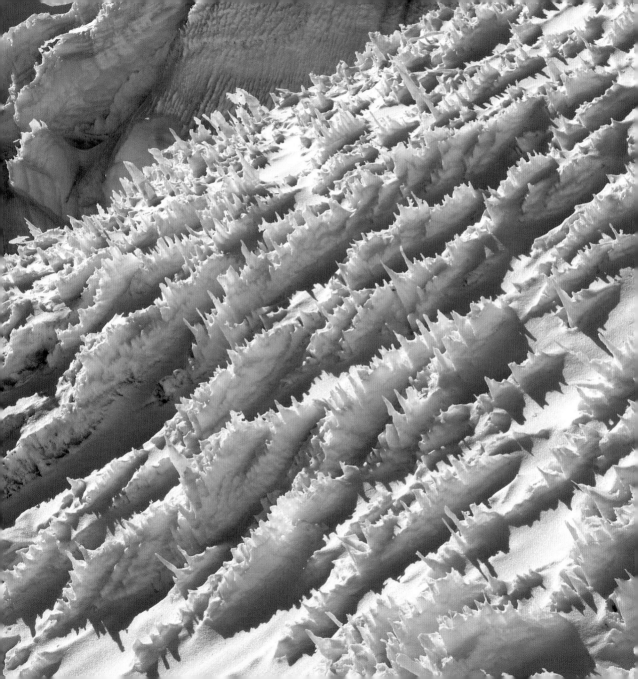

08.09.06 ▌ BOLIVIA
La Paz

Here, as in many regions of the world, urban and rural people depend on water stored in mountain glaciers and snowfields. As these "ice reservoirs" continue to shrink in the coming decades, water supply issues will get progressively more acute.

PAGES 62–63:

06.23.06 ▌ ALASKA
Columbia Bay

Bits of ice calved from the Columbia Glacier float into Columbia Bay, just west of Valdez, Alaska. Some of this ice is 300 to 500 years old, compressed snow from forgotten snowstorms.

Oil ———— is a national security risk.

OIL ADDICTION IS A NATIONAL SECURITY RISK.

DEPENDENCY ON OIL MAKES us weak. When fewer of us needed oil—and I'm not talking about back in the Stone Age, I'm talking about just a few decades ago, when a billion fewer people walked the Earth—and when Earth's stony skin had a lot more of the greasy juice available for easy pumping, oil made perfect sense. But today our dependency on oil threatens our very survival. Let's acknowledge this dependency for what it is. It is an addiction. An addiction is when you have such a compulsive need for something that your brain can barely conceive of life without it. Like all addictions, oil addiction is desperately dangerous.

Oil has been convenient, no doubt about it: Oil is deliciously protean, easily refined into a zillion different substances and moved around the globe in ships, pipelines, trucks, and airplanes. We depend on oil for transportation. We depend on petroleum-based fertilizer to feed the productivity of modern agriculture. We depend on oil to spin the fibers of our clothing. (As I write this, I'm sitting in a nylon tent, clad head to foot in four or five different forms of fleece clothing, all of it born in a refinery somewhere.) I could go on and on, but you get the idea. We are the "oil people."

Every nation on the planet is scrambling to use an ever-larger fraction of a commodity that nature isn't manufacturing anymore. The United States guzzles nearly a quarter of the 85 million barrels of oil going up in smoke on this planet every day, but in its sovereign territory holds only 2.3 percent of the world's oil endowment. That fraction was once much higher, but it has been inexorably declining since U.S. oil production peaked in 1970. Petro-blackmail is an ever-present possibility.

Oil costs us a fortune. No matter what the per barrel price might be, the United States spends a fortune on importing oil. In the first half of 2008—when the average per barrel cost of oil was $106.91—we were spending more than $2 billion per day on it, according to statistics from the U.S. Energy Information Agency. Projected over a year that petroleum would cost well over $700 billion. Texas billionaire T. Boone Pickens, once an oilman and now in the wind energy and natural gas business, calls the billions we pump into the coffers of oil-producing countries "the biggest transfer of wealth in the history of civilization." To add irony to absurdity, we borrow much of the purchase money from China and Saudi Arabia. It's a never ending cycle of self-defeat and debt. The pushers win and the junkies lose.

Consider one final cost. It's hard to know what fraction of the 2008 military budget of $696 billion should be attributed to guarding the global oil supply. Or what

fraction of the human cost of the current Iraq War—more than 4,100 Americans and nearly 100,000 Iraqis have been killed as of this writing—is part of that cost. Clearly, a major portion of both overall military spending and the Iraq War should be charged to oil defense. It looks to me like the oil we buy today is far, far too cheap. Its pricing represents a massive failure of the market system because the market isn't including the military, health—and let's not forget—environmental costs of buying and using oil. If the true expenses were added into the price at the pump, oil would likely be priced on par with the supposedly "expensive" alternative fuels such as wind and solar power. And perhaps petro fuels would cost even more than the alternatives.

How much more obvious can it get? Our addiction is overwhelming our common sense. If we break the addiction, we'll pay ourselves an enormous cash dividend by building a green energy industry in the United States. We will also be free of dependency on oil-supplying countries that do not share our political vision. I have a hard time imagining anything wrong with outcomes like that.

06.23.06 ▌ ALASKA
Columbia Glacier

Since 1984, the glacier has retreated over 10.5 miles. Glaciologists consider Columbia to be a worrisome indicator of what Greenland's gigantic tidewater glaciers might do, and may in fact already be doing.

approximately 1200 feet

06.23.06 ▮ ALASKA
Columbia Glacier

As glaciers retreat, they also get thinner. The demarcation line between the green vegetation high on the ridge and the soil and bare rock below marks the "trimline," the highest level the glacier reached in 1984. The depth of deflation is greater than the height of New York City's Empire State Building.

PAGES 70-71:

05.15.07-06.15.08 ▮ ALASKA
Columbia Glacier

A time-lapse sequence of the Columbia Glacier shows a marked retreat over eighteen months.

PAGES 72-73:

06.22.06 ▮ ALASKA
Columbia Glacier

Contrasts between clean glacial meltwater and water laden with eroded silt color these lakes on the surface of the East Fork of Columbia Glacier. Black stripes are erosional debris called "moraines."

05.15.07

10.05.07

12.31.07

01.08.08

05.09.08

05.18.08

06.15.08

THE MEANS ARE COMPLEX, THE ENDS SIMPLE.

CHILDISH FANCIES PERmeate our collective thinking. Solutions are supposed to be simple. Work is said to be worth doing only if it's fun. Stress should be avoided.

Don't shoot the messenger, but I have news from the real world: It's time to wake up. Every major accomplishment in the history of civilization—from symphonies to pyramids, from stable societies to functional families— has required hard work. (And, it must be said, some fun, too.) In times of war, leaders haven't been shy about letting their citizens know that the road ahead is rough. But during peacetime, both leaders and the led tend to shy away from confronting the big challenges.

This time it has to be different.

The solutions for revolutionizing the energy paradigm and putting a lid on climate change don't fall into the simplistic categories we prefer. Certainly, anyone listening to the mass media has become overwhelmed hearing about prospective solutions. One day, plan X is said to be the key. The next day it's plan Y. The audience is getting what the renowned *New York Times* reporter Andrew Revkin calls "journalistic whiplash." The reality is this: we need to employ a range of options that includes oil, natural gas, ethanol, coal, and nuclear power used in smarter, cleaner ways, along with proven "alternative" sources like wind and solar energy. Experimental approaches such as solar thermal, biomass, and algae may come into play. New technologies like electric cars and "smart" houses will have a huge impact. Princeton University professors Robert Socolow and Stephen Pacala have come up with the concept of "stabilization wedges," a wonderful way to understand the incremental character of all these technologies working in concert.

There is, however, one simple solution we can start now. Anyone who thinks about energy use for more than ten seconds can understand that the quickest way to "create" more energy is to stop wasting it. Vested interests have fear-mongered Americans into dreading the word "conservation." But "conservation" should become our national buzzword. It doesn't mean we must live in sackcloth and tents. It does mean we have to be more careful in our uses of energy. Most western Europeans live in great material comfort, and yet, according to the federal Energy Information Administration (USEIA), they manage a high standard of living while consuming 30 to 50 percent less energy than Americans (depending on how you crunch the numbers). We can live well, live long, live large, and live happy, even if we don't use as much energy as we do now.

Despite the political posturing and industry lobbying to the contrary, drilling more and deeper oil wells is not going to change the oil supply significantly. Because of the immutable makeup of our geologic basement, U.S. oil production peaked in 1970 at 11.3 million barrels of oil per day and has steadily declined to its current daily level of 6.8 million barrels. In short, America's oil endowment has mostly gone up in smoke already. Even if we decide to drill every square inch of the continental shelf, we have no realistic hope of adding more than a modest and short-lived additional supply to our oil endowment: According to the USEIA, technically recoverable offshore oil could amount to 59 billion barrels, enough to supply U.S. demand for 8 years. That's it. To extract this oil would mean throwing the entire coastline of the lower 48 open to drilling and turning many tranquil seaside ports and pleasant waterfronts into the kinds of petro-industrial ghettoes now found on the Gulf coast, at Alaska's Prudhoe Bay, or in Elizabeth, New Jersey.

I'll say it again: It's time to look beyond oil.

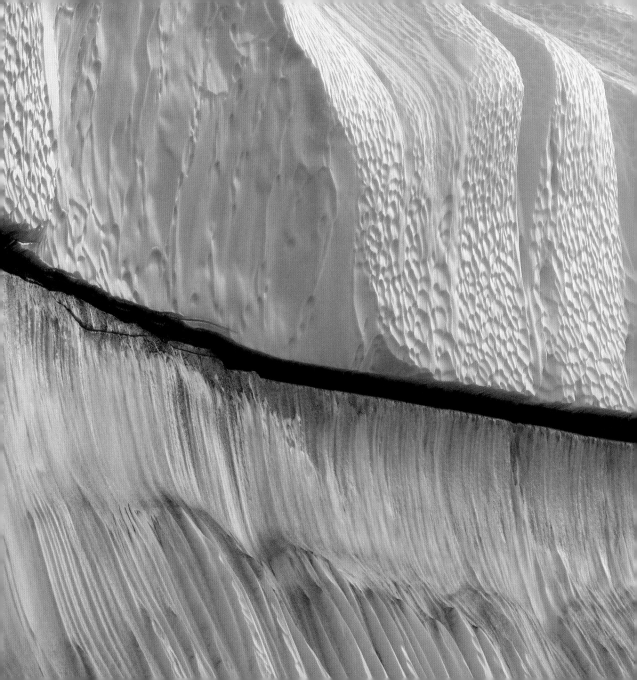

PAGES 76-77:

08.24.07 ▋GREENLAND
Disko Bay

Icebergs that have rolled over and been scalloped by waves metamorphose into fantastic shapes.

07.18.07 ▋GREENLAND
Ilulissat Isfjord

The Ilulissat Glacier puts more ice into the global ocean than any other glacier in the Northern Hemisphere. Here, tidal currents carry icebergs into Disko Bay and out to the North Atlantic Ocean. The pool of icewater seen in this shot will make its own contribution to the slow, incremental rise of global sea level.

PAGES 80-81:

05.28.08 ▋GREENLAND
Ilulissat Glacier

The largest calving incident ever caught on film, at the terminus of the glacier. The EIS team had nine cameras running during the 75-minute event.

20:46

20:48

20:50

20:51

20:55

20:58

21:03

TIPPING

Will _____ help us or hurt us?

POINTS

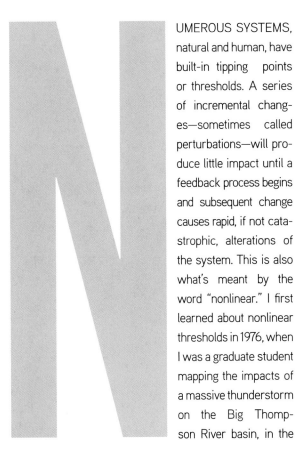

UMEROUS SYSTEMS, natural and human, have built-in tipping points or thresholds. A series of incremental changes—sometimes called perturbations—will produce little impact until a feedback process begins and subsequent change causes rapid, if not catastrophic, alterations of the system. This is also what's meant by the word "nonlinear." I first learned about nonlinear thresholds in 1976, when I was a graduate student mapping the impacts of a massive thunderstorm on the Big Thompson River basin, in the Front Range of Colorado. The deluge somehow had left little impression on most tributary canyons in the river's drainage system. But a few canyons, where a bit more rain had fallen, turned into chocolate-brown maelstroms of tumbling boulders and trees, sweeping away houses, cars, and bridges. The interaction in the system linking weather, soil, rocks, and vegetation had crossed the tipping point. The resulting flash flood killed 145 people.

Figuring out the relationship between rainfall and erosion in the Big Thompson was relatively easy after the storm. The evidence was plainly visible, etched into the muscular topography of those mountains. But before the storm, no one had a way of measuring how those relationships worked. Often, given the complexity of interactions in nature, it is difficult or impossible to know where the thresholds are until after they have been crossed. The Earth's climate system is about as complex as systems get. That's why it took so long, from the first global warming theories put forth by Charles Keeling and Roger Revelle in the 1960s, until the current decade, for the international science community to understand nonlinear climate feedbacks well enough to assert that carbon generated by humans is heating our blue sky.

Hardcore climate professionals have moved well beyond uncertainty about climate change; instead, they're debating how fast the warming is accelerating and what to do about it. There's a strong chance it's further along than we think. Mark Serreze, of the National Snow and Ice Data Center and one of the world's leading experts on the Arctic Ocean's pack ice, points out that the polar ice cap is melting much faster than was predicted by a series of forecasts produced over the past ten years by satellite pictures. He says the Arctic Ocean is likely to be ice-free by summer's end in 2030 or so, instead of 2050 or 2100, as predicted by earlier trends. If he's right, it means the region has crossed a tipping point, with one perturbation piling on top of another, in a series of ever-faster feedback loops leading to more rapid melt. The consequences to global climate of a "perturbed " Arctic Ocean changing from being a heat-reflecting surface (which it is when covered with ice) to a heat-absorbing region of ice-free water will be monumental.

Human systems cross thresholds, too, with interesting results. For example, consider the Cold War standoff between the U.S.S.R. and NATO. It was an immutable fixture of the geopolitical landscape for decades. The threat of mass annihilation hung in the air, waiting for a misstep or accident on either side. Eventually, so many people on both sides of the Iron Curtain wanted out of the system and the economic disparity between the two sides became so imbalanced that the Iron Curtain collapsed and a new international order was born.

In terms of climate change, I sense that we may be at what I call a "Berlin Wall moment," a time when obdurate and seemingly permanent barriers can collapse. Another possibility? We'll sit around on the porch of indecision not wanting to run for higher ground until the climatic storm really lets loose. By then, nature will have crossed its tipping point, and the water will be rising. We'll be caught in the climate change maelstrom and hanging on for dear life.

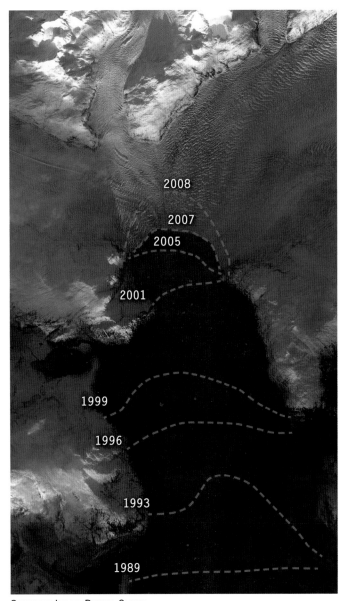

2008

2007

2005

2001

1999

1996

1993

1989

Columbia Glacier Retreat

U.S. Geological Survey scientists and EIS team member Tad Pfeffer of the University of Colorado's Institute of Arctic and Alpine Research have been monitoring the Columbia Glacier since the 1970s. The glacier became unstable due to a combination of regional climatic variations and the structural characteristics of the ice stream itself.

In the middle of the glacier's terminus, annual retreat typically equals at least a couple hundred yards, but it can be as much as one-half mile. As this book goes to press, the ice is splitting into two glaciers, one draining in from the west (left side of the frame) and another draining in from the northeast (top right side of the frame).

The extraordinarily rapid retreat is an example of how fast tidewater glaciers (i.e., a glacier that touches the ocean) can recede once local conditions become unstable. As such, Columbia's behavior is an important indicator of processes now accelerating in Greenland's tidewater glaciers.

Greenland: 2007 Melt

The larger an ice mass is, the slower it is affected by temperature changes. Thus, mountain glaciers around the world—generally smaller in mass— were the first bodies of ice to respond to the warming climate during the past few decades. Greenland has just started to react during the past two decades. Greenland's summer melting now goes to a much higher altitude and much farther inland than it did a short time ago.

Greenhouse gases, air temperatures, and summer melting are only part of the Greenland puzzle, however. Late-breaking research seems to indicate that warmer seawater around the perimeter of the subcontinent is leading to increased outflow of its tidewater glaciers. Not only does the warmer water melt the glacier but the pack ice on the sea surface is thinner and lasts a shorter time. With less pack ice to bottle up the tidewater glaciers as they attempt to flow out, the rate of ice discharge greatly increases. Greenland's future will therefore be governed by the complex interaction of air, oceans, and human impact on the Earth's natural systems.

2007
1992 and 2007
2000m line

IMAGE CREDIT: NASA.

Ilulissat Glacier

The Ilulissat Glacier once filled the long valley known as the Ilulissat Isfjord. During the early 1800s, a period referred to by glaciologists as the end of the Little Ice Age, the glacier began to retreat. Nearly two hundred years later, by 2001, the rate of retreat almost doubled. The Ilulissat Glacier had been flowing at approximately 50-60 feet per day, but after 2001 the rate accelerated to more than 120 feet per day. Many scientists believe that this speedup is the direct result of a significant rise of temperatures, both in air and in seawater, recorded since the mid-1990s.

The Ilulissat Glacier presently discharges approximately 8.4 cubic miles of ice into the ocean each year. This amount exceeds the ice discharge of all the other tidewater glaciers in the Northern Hemisphere combined.

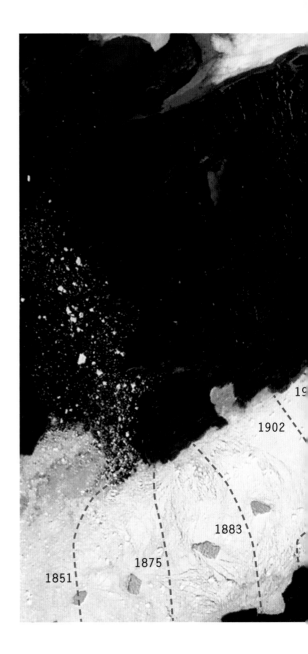

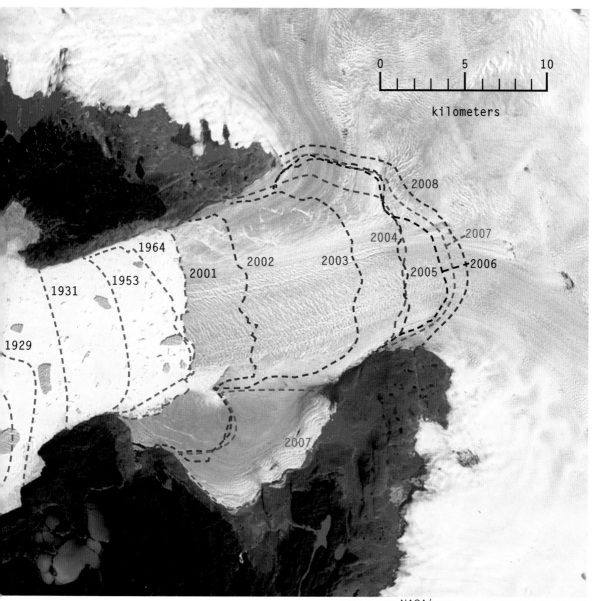

OME OF US REFLEX-ively assume that epochal, geologic-scale change has happened only in the dim past or might happen in the vague future. But that's an illusion springing from our short life spans, our limited human memory, and our cultural conditioning. Geologic processes don't, in fact, belong to some ancient age; they are happening around us all the time. They can occur in blindingly short periods of time. Floods, volcanic eruptions, tsunamis, hurricanes, earthquakes—or even climate change—the events we live through are part of the incremental process of historic change on Earth.

As climate change skeptics like to point out, nature has its own variability. But they ignore the longer narrative of climate variability spoken loud and clear by the ice cores from the Greenland and Antarctic ice sheets. This ice shows that natural climate fluctuations for nearly a million years prior to A.D. 1750 operated within a limited range dictated by natural forces. The burning of fossil fuels has added a radically new variable to that dynamic and is pushing the temperature far beyond the norms of the past million years. So profound are the impacts of burning hydrocarbon and other human activities such as deforestation and agribusiness that Paul Crutzen, winner of the 1995 Nobel Prize for chemistry, proposed renaming our era of geologic time the "Anthropocene," or human-influenced geologic period. He would place the beginning of the Anthropocene late in the 1700s, when coal began to take over from other fuels to power the industrial age. (Before Crutzen's proposal, our era was known to science as the "Holocene" and began in 8000 B.C., followed the "Pleistocene" or the Ice Ages.)

I was stunned when I first heard of it, this notion of the Anthropocene. But the more I thought of it, the more it seemed to frame an essential truth. Consider this: Until a few years ago, all of us carried a mental image of the North Pole as a place blanketed in ice, but soon, due to global warming, this will no longer be the case. Santa's workshop will have melted into the sea. Similarly, we all grew up looking into the blue sky and conceiving it as a static envelope of gas shrouding the Earth. Now we know the stasis is an illusion. How wildly improbable it all seems. Yet there it is, epochal geologic change happening right before our eyes. And what seems even more improbable? These monumental changes are produced by the trillion daily desires of a single primate species that separated itself from all the other mammals to become *Homo sapiens*.

So, dear *Homo sapiens*, we have acquired the kind of power previously reserved for rivers and tides, for oceans and mountains. With that increase in power comes an increase in obligation. We must learn to be steadfast like a mountain in the sense of living our lives with enduring strength, stability, and solidity. In our harried world, focused as it is on the ever-present now, on the individual, and on immediate financial gain, thinking long-term is not something we do much, if ever. Yet we must learn to do it well. Instead of always thinking about "feedback," we must, as the visionary architect William McDonough says, start thinking "feed forward." Short-term thinking inevitably degrades our lives and our world. The view from the mountain lets us survive in style; it shows us the way forward.

PAGES 92-93:

09.04.06 ▌ ROCKY MOUNTAINS
Sperry Glacier

Since 1850, the glacier's surface area has decreased 70-75 percent. As the ice vanished, ice-scoured bedrock, newly formed meltwater lakes, and waterfalls appeared.

09.04.06 ▌ ROCKY MOUNTAINS
Grinnell Glacier

This glacier has steadily retreated for over a century. Experts believe that it will continue retreating off the lake's surface until it is little more than a snowdrift hiding in the shadow of Mount Gould.

FUT

The ——— will judge us as either heroes or fools.

HERE ARE TIMES when the stresses of the Extreme Ice Survey, from the physical to the fiscal to the familial, make me wonder why I'm putting myself through this ordeal. But other voices, asking different questions, ultimately provide the answer.

I picture myself 30 years from now. I hear my daughters Simone and Emily (respectively 20 and 7 years old now) saying, with a certain amount of anger and exasperation: "Dad, with what everyone knew back in 2009, what the hell were they thinking? How could people not have known the climate was changing?

Why didn't anyone do anything about it?" And then they ask the personal zinger: "What were *you* doing about it?"

Hindsight can be merciless; it probably always has been. People of any given era look back in time and wonder how their predecessors could have been so dim-witted. I am amazed, for example, that people of the Middle Ages could have forgotten that inundating their streets with human fecal matter would cause disease. That was a lesson the ancient Greeks, Romans, Babylonians, and Chinese had learned long before; medieval Europeans certainly couldn't have been oblivious to the knowledge, but their lack of street sanitation makes it clear they chose to ignore it. The results were disastrous: Various forms of dysentery were common, along with outbreaks of cholera, typhoid, and other gastrointestinal illnesses. Another example: How could the British Navy have taken hundreds of years to make limes and other antiscorbutics a standard part of the sailors' diet, thus dooming tens of thousands of men to the misery of death by scurvy? The cure for scurvy—eating certain Vitamin-C rich plants—was known to European navigators as early as the 1500s. During three epic voyages of the 1770s, Captain James Cook, a nautical visionary and innovator, forced his crews to eat sauerkraut, which was later shown to be rich in Vitamin C. He lost not a single man to the disease. Yet Cook's employers, the British Admiralty, waited until 1867 before mandating through the Merchant Shipping Amendment Act, that its sailors must eat limes to stay healthy (the origin of the term "limey").

When future generations look back at the late 20th and early 21st centuries, people will be as astounded at our mule-headedness about dealing with climate change as we now are at the foibles of earlier societies. Unfortunately, the astonishing pace of climatic events, as witnessed in the ice, doesn't give us the luxury of time to be slow learners. We cannot afford to be hamstrung by ponderous social institutions or to be deluded by false information. We must make a major effort to stop climate change right now, and not think we can absolve ourselves of responsibility by educating our children to do things differently. A generation from now will be too late. The voices of the future—our grown-up children—will damn us if we are passive or indifferent.

Rise to the occasion demanded by a changing climate, and we will be heroes. Descend into collective complacency and dysfunctional denial, and we will someday be judged to have been fools. I love what Sir Edmund Hillary once said about great deeds: "People do not decide to become extraordinary. They do extraordinary things." The quality of our society and our lives will be judged by the deeds we do or don't do.

My answer to the questions my children will ask me someday is the Extreme Ice Survey. Girls, I can say to them, I did the best I could with the tools I had—photographs and video and words—to convey the immediacy and urgency of climate change. I saw it. I knew what it meant. I paid attention. I spoke as loudly as I knew how. And I did the best I could do.

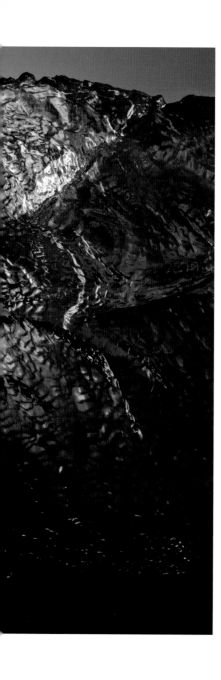

02.10.08 ▮ ICELAND
Jökulsárlón

Remnants of glaciers wash up onto the shorelines before they disappear due to warming temperatures.

PAGES 102–103:

08.10.07 ▮ GREENLAND
Store Glacier

Chunks from the Greenland Ice Sheet in the process of being flushed out to sea by Store Glacier's spring calving. These icebergs are the tangible manifestation of the process by which the ice sheet is thinning and raising sea level.

08.15.06 ▌BOLIVIA
Cordillera Real

Situated in the Bolivian Andes mountain range, this glacier has been in rapid retreat since the late 1970s, like virtually all other glaciers in the region. The lake area was once covered by ice.

ALL OF US HAVE VOICES. Voices are messages. Voices are examples. Voices can be collective, and voices can be solo. Do things differently in your personal life. At the end of this book I've given a short list of my own ideas about what you can do (page 114). Some of the changes are astonishingly simple. How many people does it take to screw in a new light bulb? How trivial is the renewable energy premium your local utility company charges? Couldn't you afford to pay those few extra dollars? Next time you must buy a car, could you afford to spend that extra 20 percent to get a hybrid? And more than a few of you could and should absorb the costs of installing photovoltaic cells on your homes. Steps like these are a more effective way of taking action in your own life than griping about the federal government's inaction.

Improving the carbon footprint of an individual life and home may seem insignificant when the rest of humanity is crashing around like a bull in a china shop. But it is the way to reshape the reality around you immediately—and that and that alone is the essence of personal empowerment. In any case, great journeys have always required a sequence of single steps. All your other actions mean influencing the behaviors of friends, family, governments, and corporations. Trying to sway your own family, let alone the whole world, is no picnic. Still you must try. I've had vitriolic debates with my extended family's climate skeptics. But oh, what a sweet day it was when suddenly, after one of my in-laws had sticker shock at the gas pump, he earnestly asked me whether he should buy a Toyota Prius like mine.

Influence your coworkers and the way your workplace does things. Let your voice be heard on the Internet; it is a great megaphone that makes a single voice heard everywhere and can join single voices into one loud voice calling for change. Let the policymakers at the local, state, and federal levels know what you want. Do the same with the companies that sell you the everyday goods of life. Do it. Do it now. There is no extra time, and there are no other people to fix climate change except us. Use your voice. Your voice creates the legacy everyone might someday celebrate us for. And remember: Despair is not an option.

PAGES 106-107: Near the Ilulissat Isfjord, Greenland, March 2008. A massive iceberg broken off the Greenland Ice Sheet, surrounded by lily pads of sea ice, in the process of breaking up at the edge of Disko Bay.

PAGES 110-111: An EIS team member provides scale in a massive landscape of crevasses on the Svinafellsjökull in Iceland, February 2008.

2007 ▍ICELAND
Remember the Miraculous

In the final analysis, the future of ice and climate change on this planet depends on how humanity deals with the nexus of energy and economics. So let me end this book with one last thought about energy supply. Paradoxical as it seems, hydrocarbon energy sources—oil, coal, and natural gas—are not ordinary fungible commodities like, say, corn kernels or pork bellies. They are a distillation of life and time in a way nothing else is. Long ago, in the dust of ancient epochs, an infinity of animals and plants were born, grew to whatever maturity was their destiny, fed, inhaled, exhaled, lived in communities, and propagated their young. Their existence—all of it based on photosynthesis—concentrated the blazing light of a trillion sunrises into their bodies, just as our own animate life does today.

The essence of those past lives and vanished sunrises is now being passed to us, transmuted into the clear liquid drizzling into our gas tanks or the flickering of a light bulb or a picture dancing across a television screen or hundreds of other daily events we never notice. We gobble up the heat and electrons in nanoseconds, and then they are gone. It is a process both wondrous and awful. Remembering the miracle of this transformation might inject some much-needed grace and humility into the engine of our consumption.

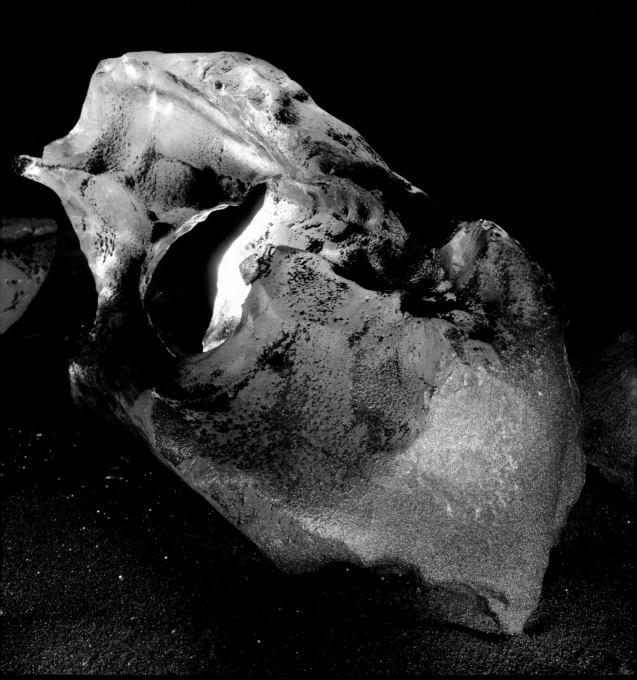

PERSONAL ACTION

The two places where you can reduce your carbon footprint the fastest? Your personal household
energy consumption is 27% of your total carbon
footprint; your transportation habits are 19%. Some
changes are small, some large. Do the best you
can. The Internet bibliography below is loaded with
more suggestions.

TO REDUCE YOUR ENERGY FOOTPRINT:
- Pay your local utility the trivial extra fee they
  charge for supporting renewable energy.
- Control your gadgets! Set your computer to
  sleep soon after you leave your desk. Turn off
  your computer, TV, stereo when not in use. Put a
  surge protector in-line between your AC outlets
  and your gadgets; turning off the entire surge
  protector eliminates standby power demand.
- Use energy-efficient appliances.
- Air-dry clothing.
- Put motion detectors on light switches.
- Change to CFL light bulbs. Watch for next generation, gel-cored bulbs.

- At home in winter: wear an extra sweater and
  warmer slippers; turn down the thermostat.
- At home in summer: don't air condition your
  rooms so cold; use a fan, swamp cooler or
  fresh air for cooling if possible.
- Install programmable thermostats.
- Use caulk or weather stripping to seal up any
  air leaks around windows and doors.
- Use ceiling fans to circulate warm air in winter,
  especially in rooms with high ceilings.

TO REDUCE YOUR TRANSPORTATION FOOTPRINT:
- Drive less, drive smarter.
- Fly less, teleconference more.
- For a personal carbon tax more meaningful
  than paying a carbon offset and more immediate than waiting for the government to act,
  invest in a hybrid car.
- Use public transport or a bike (if you're in a
  place where these solutions make sense).
- Stop drinking bottled water unless your local
  water is really awful (transporting premium

water creates a huge carbon footprint).

- Consider the possibility that you don't need to eat strawberries from Chile and kiwi from New Zealand year-round. Eat fruit more suited to the seasons of your home continent.

BIBLIOGRAPHY WEBLINKS

We Can Solve It *www.wecansolveit.org*
ConEdison *www.coned.com/thepowerofgreen/100tips.asp*
Cooperative Institute for Research in Environmental Sciences (CIRES) *www.cires.colorado.edu*
Energy Information Administration *www.eia.doe.gov*
Intergovernmental Panel on Climate Change *www.ipcc.ch*
NASA/Goddard Institute for Space Studies *www.giss.nasa.gov*
National Geographic Climate Connection *www.ngm.nationalgeographic.com/climateconnections*
National Religious Partnership for the Environment *www.nrpe.org*
National Renewable Energy Lab *www.nrel.gov*
National Resources Defense Council *www.nrdc.org*
National Snow and Ice Data Center (NSIDC) *www.nsidc.org*
United Nations Environment Programme *www.unep.org*

A FEW MORE FACTS

- Since global temperature recording began in 1850, the ten hottest years on record, in order from hottest to less hot, have been: 2005, 1998, 2002, 2003, 2007, 2006, 2004, 2001, 1997, 1995.
- Northern Hemisphere temperatures in the last half of the 20th century were warmer than any other 50-year period of the previous 500 years; it has probably been the warmest half-century in the past 13,000 years.
- Producing one megawatt hour of power from coal releases 2,250 pounds of carbon dioxide; oil releases 1,700 pounds; natural gas releases 1,125 pounds.
- The U.S. consumes approximately 20 billion barrels of oil per day, or 25% of daily global production. The U.S. has 4.5 % of the world's population.

EIS requires the physical effort, expertise, financial backing and/or emotional support of more people, more often, more consistently, than any other undertaking in my thirty-year career. Seeing all these names consolidated below gives me an awestruck and humble appreciation of this. Available space limits my ability to describe the unique and important contributions of each person and organization, but I hope that everyone mentioned understands that my gratitude is not less for its brevity.

My dearest family—my wife Suzanne and my daughters Simone and Emily—are the absolutely vital foundation on which EIS is built; they are the hearth to which this Odysseus happily returns after innumerable absences. For their cheerful, patient support I offer my deepest thanks. My parents, James and Alvina, share in the EIS saga in many, many ways, as my brothers Stephen and Michael have.

The National Geographic Society gave me the crucial first assignment from which EIS grew, then became vital in making EIS itself happen. Rebecca Martin and John Francis at the Expeditions Council, and Chris Johns, Dave Griffin, and Dennis Dimick at *National Geographic* magazine (supported by Barbara Moffett) led the charge. The executive staff—Terry Garcia, John Q. Griffin, and Tim Kelly—have steadily embraced our effort. *National Geographic Adventure*, led by John Rasmus and Sabine Meyer, and supported in the field by Pete McBride and Michael Shnayerson, gave us a wonderful forum. The television division—Steve Reverand, John Bredar, Marianne Culpepper, and Nancy Donnelly—partnered with Far West Film's Noel Dockstader and Quinn Kanaly, and NOVA's Paula Apsell, to produce an excellent look at the ice. My incredibly hardworking editor Bronwen Latimer and art director Melissa Farris, along with the book division's Kevin Mulroy and Leah Bendavid-Val, created a terrific opportunity for EIS imagery to nestle between the covers of this book.

Many at Nikon USA—Joe Ventura, Anna-Marie Bakker, Bill Pekala, Steve Heiner, Diane Bachman, and Ron Taniwaki—make sure that EIS has the equipment we need for watching the glaciers change. NASA and its program officers, Seelye Martin and Ming-Ying Wei, and Jason Buenning, Robin Abbott, and Mark Begnaud at CH2MHill, help make our Greenland work happen, while the National Science Foundation's Mike Ellis keeps the Alaskan work alive.

Jason Box of Ohio State University and Tad Pfeffer of the University of Colorado are friends and allies, confidantes and research partners; their knowledge, their funding, their field expertise are essential. My field assistants, Adam LeWinter, Jeff Orlowski, and Svavar Jónatonsson, bring great passion and energy to EIS (Jeff also devotes countless hours to his superb video shooting and editing). The EIS studio manager, Sport, keeps the frenetic wheel of logistics turning smoothly. Michael Aisner infuses

his marketing wizardry into many different issues. Dan Fagre, of the U.S. Geological Survey, anchors our Glacier National Park work in Montana, as Eran Hood, Dave Sauer and Nick Korzen of the University of Southeast Alaska do on Mendenhall Glacier. David Finnegan, of the Cold Regions Research and Engineering Laboratory, collaborates on Hubbard Glacier. Greg Marshall, Corey Jaskolski, and the rest of the National Geographic Remote Imaging Department, and Neil Humphrey of the University of Wyoming, play critical roles in moving technical issues forward. Michael Brown, Ben Phelan, David Breashears, and Dave Ruddick put their tremendous skills as cinematographers to brilliant use during long days in difficult conditions. The NBC team of Brian Williams, Anne Thompson, and Mario Garcia did a spectacular story for the Nightly News.

The Former Vice President and Nobel Laureate Albert A. Gore, Jr. and Tipper Gore gave important validation of EIS and helpful information about British Columbia glaciers. Denver Mayor John Hickenlooper and Denver International Airport Director Kim Day opened the door to fantastic exposure for the project. Sarah Das, Ian Howat, Ian Joughin, and Slawek Tulaczyk are great companions out on the ice. Others who help out on field work are Yushin Ahn, Alberto Behar, Laurie Craig, Rollo Garabotti, Olivier Greber, Didier Lavigne, Luc Moreau, Shad O'Neel, and an Italian character in Greenland known only as Silver. The stateside support team also includes Roy McCutchen, Jen Jones, Mike Harrison, and John Botkin of Photocraft Laboratories, and Scott Roche and Ed Kaufman at Coupe Studios (both firms are in Boulder (CO)); Sherri Leopard, Brendan Hemp and Martin Walaszek at Leopard Communications in Broomfield (CO); Laura Caruso, Lucien Foehr, Daniel Goldhaber, Blake Gordon, Brad Kahland, Jeff Kent, Eric Nance, Barb Shively, Renee Thompson, Eric Turner, Matt Vellone, and Nick Weinstock. Chuck Herring, Amy Opperman and the terrific staff at Digi-

tal Globe have provided priceless satellite imagery.

Scientific advice from people outside the EIS team is vital, particularly the seminal encouragement and continuing counsel of Konrad Steffen, director of the Cooperative Institute for Research in Environmental Sciences at the University of Colorado. I'm thankful also for the guidance and energetic collaboration from Helgi Bjornsson, Mark Fahnestock, Bernard Francou, Einar and Gulli Gunnlaugsson, Meredith Nettles, Oddur Sigurdsson, Martin Truffer, and Christian Vincent.

Richard Goldman and Amy Lyons of the Richard and Rhoda Goldman Fund; Brooks Fisher, Andrea and Tim Fisher and all the members of their extended family; Rebecca DiDomenico and Stephen Perry, Edith Eddy and everyone on the Compton Foundation board; Shai and Nili Agassi, Annie Griffiths Belt, Rich Clarkson, Wade Davis, Sylvia Earle, Jesse and Betsy Fink, David Friend, Martin Goebel, David Holbrooke and Telluride Mountainfilm, Ginny

Jordan, Charlie Knowles, Bob and Gail Loveman, Cristina Mittermeier and the International League of Conservation Photographers, Gib and Susan Myers, Chris Rainier, Bill Unger; the Aspen Institute's Kitty Boone, Elliot Gerson, and David Monsma: each and everyone of you has my deepest gratitude for teaming with EIS in all the different ways you have. Katie Ramage and Letitia Ferrier at The North Face; Kris Brunngraber at Bogen Imaging; Michelle Kerkvoordian at Pelican Products; and Craig Mintzlaff at Rudy Project have made sure we were well-outfitted. The wonderful endorsements of the Rowell Award committee members, particularly Conrad Anker, Frans Lanting, Corey Rich, and Brian Thyssel, and that of the North America Nature Photography Association were tremendously appreciated.

And finally, to this, the sun's third stone, the home that has put up with so much for so long: you are a great home, and for that we thank you, all six billion of us. We will try to do no more harm.

JAMES BALOG is the author of seven books, including *Tree: A New Vision of the American Forest* and *Survivors: A New Vision of Endangered Wildlife,* which were hailed as major breakthroughs in nature photography. His images are regularly published in magazines like *The New Yorker, American Photo, Vanity Fair, Sierra, Audubon* and *Outside* and he is a contributing editor to *National Geographic Adventure.* He is the first photographer ever commissioned to create a series of stamps for the U.S. Postal Service. Awarded the Leica Medal of Excellence and the premier awards for both nature and science photography at World Press Photo in Amsterdam, Balog has exhibited photos at more than a hundred museums and galleries from Paris to Los Angeles. The documentary film "A Redwood Grows in Brooklyn" explores his thoughts about art, nature and perception. He lives on a Rocky Mountain ridge above Boulder, Colorado.

EXTREME ICE NOW
JAMES BALOG

PUBLISHED BY THE NATIONAL GEOGRAPHIC SOCIETY

John M. Fahey, Jr., *President and Chief Executive Officer*
Gilbert M. Grosvenor, *Chairman of the Board*
Tim T. Kelly, *President, Global Media Group*
John Q. Griffin, *President, Publishing*
Nina D. Hoffman, *Executive Vice President;*
　　　　　President, Book Publishing Group

PREPARED BY THE BOOK DIVISION

Kevin Mulroy, *Senior Vice President and Publisher*
Leah Bendavid-Val, *Director of Photography Publishing & Illustrations*
Marianne R. Koszorus, *Director of Design*
Barbara Brownell Grogan, *Executive Editor*
Elizabeth Newhouse, *Director of Travel Publishing*

STAFF FOR THIS BOOK

Bronwen Latimer, *Project and Illustrations Editor*
Melissa Farris, *Art Director*
Karin Kinney, *Text Editor*
Sport, *Editorial Coordinator*
Robert Waymouth, *Illustrations Specialist*
Jennifer A. Thornton, *Managing Editor*
Gary Colbert, *Production Director*
Meredith C. Wilcox, *Administrative Director, Illustrations*

MANUFACTURING AND QUALITY MANAGEMENT

Christopher A. Liedel, *Chief Financial Officer*
Phillip L. Schlosser, *Vice President*
Chris Brown, *Technical Director*
Nicole Elliott, *Manager*
Monika D. Lynde, *Manager*
Rachel Faulise, *Manager*

Founded in 1888, the National Geographic Society is one of the largest nonprofit scientific and educational organizations in the world. It reaches more than 285 million people worldwide each month through its official journal, NATIONAL GEOGRAPHIC, and its four other magazines; the National Geographic Channel; television documentaries; radio programs; films; books; videos and DVDs; maps; and interactive media. National Geographic has funded more than 8,000 scientific research projects and supports an education program combating geographic illiteracy.

For more information, please call 1-800-NGS LINE (647-5463) or write to the following address:

National Geographic Society
1145 17th Street N.W.
Washington, D.C. 20036-4688 U.S.A.
Visit us online at www.nationalgeographic.com

For information about special discounts for bulk purchases, please contact National Geographic Books Special Sales: ngspecsales@ngs.org

For rights or permissions inquiries, please contact National Geographic Books Subsidiary Rights: ngbookrights@ngs.org

Focal Point is an imprint of National Geographic Books
ISBN: 978-1-4262-0401-2
Printed in China